Kyongyob Min

What is the Biological Variability ?

Kyongyob Min

What is the Biological Variability ?

Biological varaibility, Stochastic optimal control theory and Shrödinger's wave equation

LAP LAMBERT Academic Publishing

Impressum / Imprint

Bibliografische Information der Deutschen Nationalbibliothek: Die Deutsche Nationalbibliothek verzeichnet diese Publikation in der Deutschen Nationalbibliografie; detaillierte bibliografische Daten sind im Internet über http://dnb.d-nb.de abrufbar.
Alle in diesem Buch genannten Marken und Produktnamen unterliegen warenzeichen-, marken- oder patentrechtlichem Schutz bzw. sind Warenzeichen oder eingetragene Warenzeichen der jeweiligen Inhaber. Die Wiedergabe von Marken, Produktnamen, Gebrauchsnamen, Handelsnamen, Warenbezeichnungen u.s.w. in diesem Werk berechtigt auch ohne besondere Kennzeichnung nicht zu der Annahme, dass solche Namen im Sinne der Warenzeichen- und Markenschutzgesetzgebung als frei zu betrachten wären und daher von jedermann benutzt werden dürften.

Bibliographic information published by the Deutsche Nationalbibliothek: The Deutsche Nationalbibliothek lists this publication in the Deutsche Nationalbibliografie; detailed bibliographic data are available in the Internet at http://dnb.d-nb.de.
Any brand names and product names mentioned in this book are subject to trademark, brand or patent protection and are trademarks or registered trademarks of their respective holders. The use of brand names, product names, common names, trade names, product descriptions etc. even without a particular marking in this works is in no way to be construed to mean that such names may be regarded as unrestricted in respect of trademark and brand protection legislation and could thus be used by anyone.

Coverbild / Cover image: www.ingimage.com

Verlag / Publisher:
LAP LAMBERT Academic Publishing
ist ein Imprint der / is a trademark of
OmniScriptum GmbH & Co. KG
Heinrich-Böcking-Str. 6-8, 66121 Saarbrücken, Deutschland / Germany
Email: info@lap-publishing.com

Herstellung: siehe letzte Seite /
Printed at: see last page
ISBN: 978-3-659-55694-4

What is the Biological Variability?

Biological Variability, Stochastic Optimal Control Theory and Shrödinger's wave equation

Kyongyob Min, MD & PhD
Respiratory Division, Department of Internal Medicine, Itami City Hospital1-100 Koyaike, Itami, Hyogo,
JapanEmail: in1007@poh.osaka-med.ac.jp

CONTENTS

Chapter 1

What is the Biological Variability? - as an Introduction

Classical physiology is grounded on the principle of homeostasis, in which regulatory mechanisms act to reduce variability and to maintain a steady state. Cherniack et al. applied a system engineering approach to the control of respiration, describing a controller (brain stem respiratory pattern generator), sensors (chemo- and mechanoreceptors), and a plant (airways, chest wall, muscles, and pulmonary tissue). With this model, fluctuations are often dismissed as "noise" of little or no significance. However, biological processes in the body provide an endless and astounding source of complexity. For example, Fadel et al. measured interbreath intervals (IBIs) and tidal volumes of human adults, and demonstrated that spontaneous breathing was noisy (Figure 1A). Frey et al. measured IBIs for both a preterm baby at 39 and 61 weeks of postconceptional age, and showed that the baby's breathing pattern was highly irregular at 39 weeks and the fluctuations were significantly reduced by 61 weeks (Figure 1B).

The complexity associated with biological processes would produce significant fluctuations in physiological parameters at the organ level as a consequence of the biological system being held under a nonequilibrium steady-state condition (a "homeokinetic" condition). Since many systems in nature, including respiration, operate away from an equilibrium point, the importance of taking fluctuations into account was well known from early models of the respiratory control mechanism.

The fluctuations in physiological parameters are not simply attributable to random noise superimposed on regular processes, but often show signs of complex behavior.

Examining how the variable follows itself in time, it becomes obvious that physiological signals show temporal correlations. One can use the power spectrum, which displays the frequency content and any periodicity to characterize the correlations. If the correlations span several orders of magnitude, the signal is said to have long-range correlations. For example, normal heart rate displays significant variability with long-range correlations and breakdown of this behavior signifies disease of emergent exacerbation. Thus, researchers began to approaches to analyze various noisy data recordings.

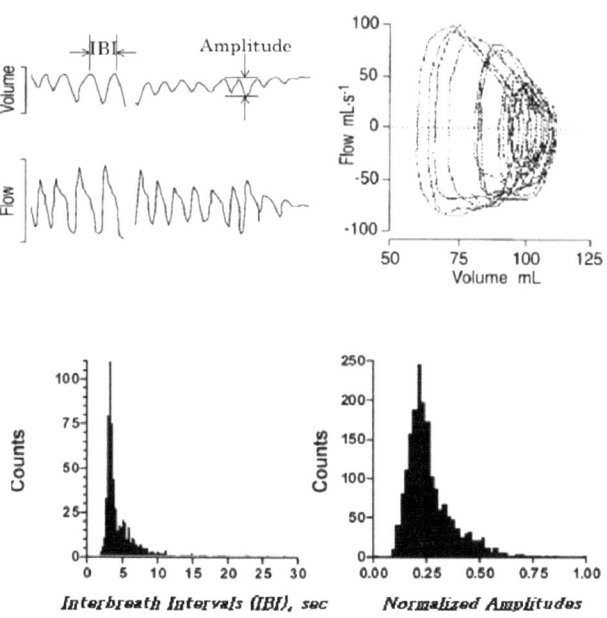

Figure 1A.   **Records of Noisy Breathing with Computing Histogram**
(modified from Fadel PJ et al, 2004)

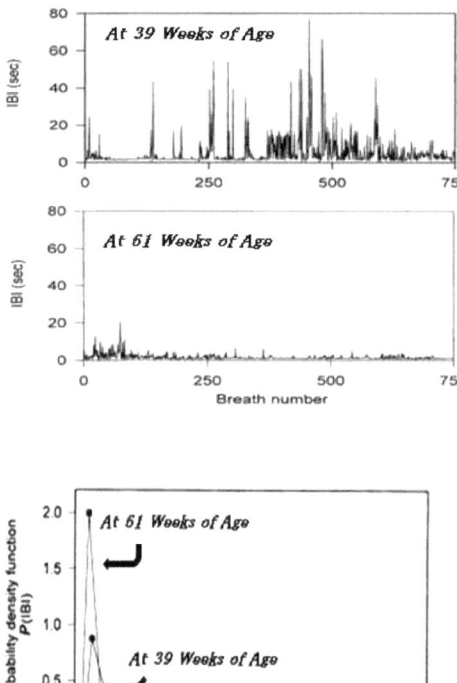

Figure 1B. Comparison of Porbability Density Distributions Between infants at age of 39 weeks and at age of 61 weeks (modified from Frey U et al. 1998)

A widely used method to quantify correlations among noisy data is called the detrended fluctuation analysis (DFA). Then, quite surprisingly, DFA on respiratory variables has suggested no correlations in the data as if the fluctuations were totally random. Fluctuations in respiratory variables should be characterized in another way. The extent of fluctuations can be characterized from temporal variations by computing the histogram or probability distribution of the variable. Even if DFA on data showed no temporal correlations, one can characterize the fluctuations by the

6

probability distribution, which would signify the biological system being in a nonequilibrium steady-state or "homeokinetic" condition. Often, the probability distribution of physiological signals displays a long tail that is best visualized on a double logarithmic scale. If the tail decays linearly on the log-log-graph, the distribution is said to have power law tail. Since the linear decay on the log-log graph has no characteristic peaks or features at any particular scale, the power law is said to show a scale-free or self-similar ("fractal") behavior.

For constructing realistic models of control mechanisms with biological variability in spontaneous breathing, one is faced with the problem of finding suitable ways to characterize them. A characteristic feature of fluctuations is the impossibility of precisely predicting their future values, and thus some researchers have tried to use statistical concepts to model fluctuations. From this statistical viewpoint, Frey et al. and Suki have suggested three points on noisy biological variables: (1) the fluctuations obey their own probability distribution, (2) irregular fluctuations can carry information through the probability distribution, and (3) the probability distribution may be sensitive to physiological or pathological changes. Thus, I would like to define the biological variability by both a set of random variables and corresponding specific probability distribution.

References

Fadel PJ, Barman SM, Phillips SW, and Gebber GL. Fractal fluctuations in human respiration. *2004 J Appl Physiol 97:*2056-2064

Frey U, Silverman M, Barrabasi AL, Suki B. Irregularities and power law distributions in the breathing pattern in preterm and term infants. *1998 J Appl Physiol 85*: 789-797

Frey U, Maksym and Suki B. Temporal complexity in clinical manifestations of lung disease. *2011 J Appl Physiol 110:* 1723-1731

Jacono FJ and Dick TE. Variability, measuring the spice of life. *2011 J Appl Physiol 111*: 351-352

Macklem PT. Emergent phenomena and the secrets of life. *2008 J Appl Physiol 104*:1844-1846

Kantelhardt JW, Koscielny-Bunde E, Rengo HHA, Havlin S, Bunde A. Detecting long-range correlations with detrended fluctuation analysis. *2001 Physica A 295*:441-454

Nemati S, Edwards BA, Sands SA, Berger PJ, Wellman A, Verghese GC, Malhorta A, Butler JP. Model-based characterization of ventilator staqbility using spontaneous breathing. *2011 J Appl Physiol 111*:55-67

Peng CK, Havlin S, Stanley HE, Goldberger AL. Fractal mechanisms and heart rate dynamics. Long-range correlations and their breakdown with disese. *1995 J Electrocardiol 28, Suppl*: 59-65

Suki B. Fluctuations and power laws in pulmonary physiology. *2002 Am J Respir Clrit Care Med 166*:1233-137

Suki B. In search of complexity. *2010J Appl Physiol 109*: 1571–1572

Chapter 2: Differentiable Stochastic Processes

2.1 Fluctuations as a Sequence of Random Variables

A characteristic feature of fluctuations is the impossibility of precisely predicting their values. A successful attempt is to model a disturbance as a sequence of random variables or a stochastic process. A stochastic process can be defined as a family of random variables $\{X(t), t = t_0, t_0 + 1, \ldots\}$. It is possible to assume that the random variables $X(t)$ represent values on the real line or in an n-dimensional Euclidean space. A stochastic process is a function of two arguments $\{X(t, \omega)\}$, where ω belongs to the sample space Ω. For fixed t, $X(t, \cdot)$ is a random variable and for fixed ω, $X(\cdot, \omega)$ is a function of time which is called a sample function or a trajectory. The trajectories can be regarded as elements of the sample function space Ω. For ordinary random variables whose sample function spaces are Euclidean spaces, probability measures can be assigned by ordinary distribution functions and denoted by P.

Let us assign a probability function to the multidimensional random variable $\{X(t_1), X(t_2), \cdots, X(t_3)\}$ for any k and arbitrary time t_j with a distribution function F as follows,

$$F(\xi_1, \xi_2, \cdots, \xi_k; t_1, t_2, \cdots, t_k) = P\{X(t_1) \leq \xi_1, X(t_2) \leq \xi_2, \cdots, X(t_k) \leq \xi_k\} \cdots (2.1.1)$$

which satisfies the conditions of symmetry in all pairs (ξ_j, t_j) and consistency. The consistency condition is expressed by

$$F(\xi_1, \xi_2, \cdots, \xi_k; t_1, t_2, \cdots, t_k) = \lim_{t_k \to \infty} F(\xi_1, \xi_2, \cdots, \xi_k; t_1, t_2, \cdots, t_k) \cdots (2.1.2)$$

Thus, the mean value of a stochastic process m(t) is defined by use of the probability distribution density $dF(\xi, t)$ as follows,

$$m(t) = \int_{-\infty}^{+\infty} \xi dF(\xi, t) = E[X(t)] \quad \cdots (2.1.3)$$

The symbol E[ ] denotes expectation, that is, integration with respect to the measure P. The covariance of $X(s)$ and $X(t)$ are also given by

$$r(s,t) = cov[X(s), X(t)] = E[\{X(s) - m(s)\}\{X(t) - m(t)\}] \quad \cdots (2.1.4)$$

When both the mean value function m(t) and the covariance r(s,t) exist, the stochastic process is said to be of second order.

2.2. A Wiener Process and a Markov Process

Let us consider the stochastic process of second order $\{X(t_j), j = 1,2,3,\ldots,k\}$, and $t_1 < t_2 < t_2 < \cdots < t_k$. When the set elements $\{X(t_k) - X(t_{k-1}), X(t_{k-1}) - X(t_{k-2}), \ldots, X(t_2) - X(t_1), X(t_1)\}$ are mutually independent, the process is called a process with independent increments. If the variables are only uncorrelated, the process $\{X(t)\}$ is called a process with uncorrelated or orthogonal increments. A Wiener process is one with orthogonal increments defined by the following conditions: 1) $X(0)=0$, 2) $X(t)$ is normal, 3) $m(t)=0$ for all $t>0$, and 4) the process has independent stationary increments. Since a Wiener process has independent stationary increments and $X(0)=0$, the variance of the process is $var[X(t)] = \sigma_X^2(t) = ct$, and the covariance of the process is $r(s,t)=c$ x (the minimal difference between t and s), where the parameter c is called the variance parameter.

A stochastic process $\{X(t)\}$ is called a Markov process if

$$P\{X(t) \leq \xi | X(t_1), X(t_2), \cdots, X(t_k)\} = P\{X(t) \leq \xi | X(t_k)\} \quad \cdots (2.2.1)$$

where $P\{\cdot | X(t_k)\}$ denotes the conditional probability given $X(t_k)$. When the initial probability distribution $F(\xi_1; t_1) = P\{X(t_1) \leq \xi_1\}$ and the transitional probability distributions $F(\xi_t, t | \xi_s, s) = P\{X(t) \leq \xi_t | X(s) = \xi_s\}$ are given, the distribution function of the trajectory $\{X(t_1), X(t_2), \cdots, X(t_k)\}$ is given by the Bayes' rule as follows,

$$F(\xi_1, \xi_2, \cdots, \xi_k; t_1, t_2, \cdots, t_k)$$

$$= F(\xi_k, t_k | \xi_{k-1}, t_{k-1}) \cdots F(\xi_2, t_2 | \xi_1, t_1) F(\xi_1; t_1) \cdots (2.2.2)$$

(2.2.2) shows that a Markov process is defined by both the initial probability distribution and the transition probabilities.

2.3. Stochastic State Models

State models, i.e., systems of first order difference or differential equations, are very convenient for the analysis of systems. An extension of this concept to stochastic state models requires that the probability distribution of the state variable x at future time should be uniquely determined by the actual value of the state. If $X(t+1)$ is a random variable which depends on the state variable x at the time t

$$X(t+1) - X(t) = b(x, t) + \varepsilon(x, t) \cdots (2.3.1)$$

where b(x,t) and $\varepsilon(x, t)$ are the conditional mean of X(t+1) and a random variable given the state variable x at the time t. When the model (2.3.1) is a Markov process, the conditional distribution of $\varepsilon(x, t)$ given x is normal and the stochastic variable $\varepsilon(x, t)$ can always be normalized by its variance σ^2 through a Wiener process w(t) with unit variance parameter,

$$\varepsilon(x, t) = \sigma w(t) \cdots (2.3.2)$$

2.4 Stochastic Differential Equations of State Models

Starting with the difference

$$X(t+h) - X(t) = b(x, t)h + o(h^2) \cdots (2.4.1)$$

where the term $o(h^2)$ denotes the omit terms of higher order than 2. One can easily obtain a stochastic difference equation by adding a disturbance $\varepsilon(x, t)$,

$$X(t + h) - X(t) = b(x, t)h + [\varepsilon(x, t + h) - \varepsilon(x, t)] + o(h^2) \cdots (2.4.2)$$

When the disturbance $\{\varepsilon(x, t)\}$ is a Markov process with independent increments, the conditional distribution of $[\varepsilon(x, t + h) - \varepsilon(x, t)]$ given x is normal. Hence

$$\varepsilon(x, t + h) - \varepsilon(x, t) = \sigma[w(t + h) - w(t)] \cdots (2.4.3)$$

where $\{w(t)\}$ is a Wiener process with unit variance parameter. Thus, the stochastic state model is obtained for the stochastic process $\{X(t)\}$

$$X(t + h) - X(t) = b(x, t)h + \sigma[w(t + h) - w(t)] + o(h^2) \cdots (2.4.4)$$

Therefore, the expectation $E[X(t+h)-X(t)]$ and the variance $var[X(t+h)-X(t)]$ are obtained as (2.4.5) and (2.4.6) respectively,

$$E[X(t + h) - X(t)] = b(x, t)h + o(h^2) \cdots (2.4.5)$$

$$\text{var}\, [X(t + h) - X(t)] = \sigma^2 E[w(t + h) - w(t)]^2 + o(h^2)$$

$$= h\sigma^2 + o(h^2) \cdots (2.4.6)$$

Then, let h go to zero in (2.4.4) and one obtains the following formal expression (2.4.7)

$$X(t + dt) - X(t) = dX(t) = b(x, t)dt + \sigma dw(t) \cdots (2.4.7)$$

which is called a stochastic differential equation. The function of $b(x,t)$ is called a forward drift function of the state x at the time t.

The stochastic differential (2.4.7) is defined as the limit of (2.4.4). However, another expression is possible for $dX(t)$ as follows,

$$dX(t) = X(t) - X(t - h) \cdots (2.4.8)$$

The difference $[w(t) - w(t - h)] = dw^*(t)$ is not dependent on $X(t-h)$ but on $X(t)$, and the variance of $dw^*(t)$ is $\text{var}[dw^*(t)] = -dt$. Then, another stochastic differential equation is possible as follows,

$$X(\text{t}) - X(\text{t} - \text{dt}) = dX(\text{t}) = \text{b}^*(x, \text{t})dt + \sigma dw^*(\text{t}) \cdots (2.4.9)$$

where b*(x,t) is a backward drift function of the stochastic process given x at the time t.

References

Astroem KJ. Introduction to Stochastic Control Theory. *1970 Dover Books on Electrical Engineering, NY,*:Chapters 1, 2, and 3

Chapter 3: State Variables in Noisy Breathing

3.1 Fractal Geometry of Bronchial Tree

Since the late 1980s, high-resolution CT (HRCT) has been established as an indispensable tool in the evaluation of patients with suspected diffuse pulmonary disease. It is commonly used in clinical practice to detect and accurately characterize a variety of lung abnormalities. HRCT provides a tool capable of accurately demonstrating gross anatomy and accurately characterizing abnormal findings. Recently, it has been proposed to establish the center for pulmonary functional imaging in Harvard medical center with a mission of developing and utilizing advanced imaging technologies including HRCT to measure regional and temporal changes of pulmonary functions including ventilation, perfusion, motion and gas exchange. The accurate interpretation of HRCT images requires a detailed understanding of normal lung anatomy and architectural properties of anatomical components from a view of geometry.

Biologists with an interest in the quantitative aspects of biological branching structures, including bronchial trees, must perform laborious procedures to obtain · good data and as a consequence, have implemented the properties of fractal (scale-independent self-similarity) geometry. Firstly, there is a power relationship between the diameters at a bifurcation $(r_1, r_2,$ and $r_3)$ (Figure 2A),

$$r_1^n = r_2^n + r_3^n \cdots (3.1.1)$$

Secondarily, the most important step to describe the quantitative aspects of biological branching trees is to introduce ordering systems of classifying each branch of tree structure. In the long sequence of papers on lung airways by Horsfield et al., the ordering method originally adopted is related to that proposed by Shreve, except that the edges of order j and order k come together at a vertex, and the third edge is assigned not to j + k but to one order greater than the greater of j and k, or to j + 1 if j = k. Horsfield's ordering scheme is centripetal and topological and includes only edges (Figure 2B).

Horton drew attention to a number of empirical regularities, usually now known as Horton's laws. The work of Horton encouraged several investigators to find regularities corresponding to geometrical self-similarity or fractal properties, which denotes geometrical self-similarity of the mean length of a branch. A Horton's law is expressed as follows,

$$\frac{\bar{L}_{j+1}}{\bar{L}_j} = 2^\lambda \text{ (constant) } \cdots (3.1.2)$$

where $\bar{L}_j$ and $\bar{L}_{j+1}$ denote the mean length of a branch of Horsfield's order j and j+1, respectively. Several investigators, such as Horsfield and Cumming, Raabe et al., and Horsfield and Thurlbeck, reported Horton's ratio for the bronchial tree as ranging between 1.33 and 1.92. We used $\lambda = 0.5$ for the bronchial tree of humans. According to measurements of the length-radius relationship of various arterial branching by Suwa and Takahashi, additional relationships have been reported as follows:

$$\frac{\bar{L}_j}{\bar{r}_j^i} = h \text{ (constant) } \cdots (3.1.3)$$

$$i + n = 4 \qquad \cdots (3.1.4)$$

where $\bar{L}_j$ and $\bar{r}_j$ denote the mean length and diameter of the bronchial branch of order j, respectively. Thus, the bronchial tree is a fractal tree characterized by the set of power laws.

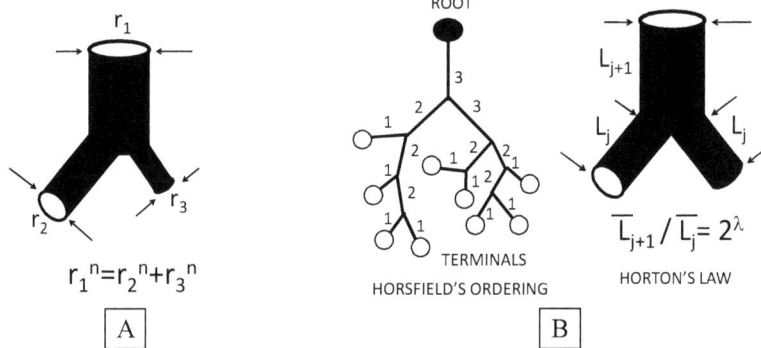

Figure 2: Power laws between diameters at a bifurcation: Flow travels through the diameter of larger parent branch (r_1) connected to two smaller diameter daughter branches (r_2 and r_3) with flow adhering to a local power-law scaling relationship. The n is the junctional exponent. Data of bronchial or arterial trees were summarized by Suwa et al. and expressed by power of n=2.7. B: Horsfield's ordering system and Horsfield's version of Horton's law (see text in detail) (modified from Min K et al. 2012)

3. 2 Shifted phasic contractions of muscular bundles in Lobular bronchiole

A simple physical law that governs laminar flows of Newtonian fluids of a viscosity constant η through non-distensible circular tubes of L in length states that a pressure drop ΔP to flow Q is expressed by an equation as follows, $\Delta P = 8\eta LQ/\pi r^4$. Based on the power laws in the bronchial tree (according to the equations (1.1) and (3.1.3)), the length of a branch L_j and the flow Q_j are expressed by a corresponding diameter of r_j as $L_j = hr_j^4$ and $Q_j = qr_j^n$, where h and q are constant parameters. Thus, the pressure drop ΔP along a branch is constant or even because of $i + n = 4$ as follows,

$$\Delta P = \frac{8\eta L_j}{\pi r^4} Q_j = \left(\frac{8\eta hq}{\pi}\right) r^{i+n-4} = \left(\frac{8\eta hq}{\pi}\right) \text{(constant)} \cdots (3.2.1)$$

Thick bundles of smooth muscle have been found in the wall of the lobular bronchioles, which regulates ventilation of the corresponding lobule over time. Krahl directly observed asynchronous lobular ventilation and perfusion through a window on the ribcage of rabbits in vivo, and reported that shifted phasic lobular perfusion and ventilation were seen during spontaneous breathing, and disappeared after surgical vagonectomy.

3.3 Miller's secondary pulmonary lobules

HRCT images of pulmonary parenchyma have revealed the anatomical and functional importance of Miller's secondary pulmonary lobule, which refers to the smallest unit of lung structure marginated by connective tissue septa. Secondary pulmonary lobules are irregularly polyhedral in shape and vary in size, measuring from 1 to 2.5cm in diameter in most locations. In one study, the average diameter of secondary pulmonary lobules measured in several adults ranged from 11 to 17mm. Airways, pulmonary arteries and veins, lymphatics, and the various components of pulmonary interstitium are all represented at the level of the pulmonary lobule (Figure 3A and B). The pulmonary lobe is composed of lobular polyhedrons (Miller's secondary lobules), each of which is supplied by a single bronchiole adjoined to an edge of a polyhedron (Figure 4). The bronchial tree is located in the adventitia that borders the lobular polyhedrons. Thus, the airway-parenchymal interdependence exists through interrelation between the lobular bronchiole and the corresponding lobular polyhedron.

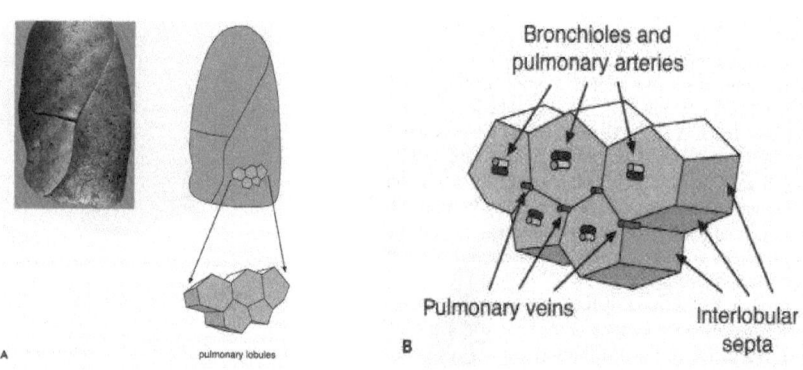

Figure 3: Pulmonary lobular anatomy
A: Pulmonary lobules that are irregulary polyhedral or conical in shape are often visible on the surface of the lung, as shown in this diagram of five lobules visible on the posterior surface of the right lung. B: Lobules are supplied by a small bronchiolar and pulmonary artery branches, which are central in location. They are variably marginated by connective tissue interlobular septa that containe pulmonary vein and lymphatic branches. (Specieme photograph courtesy of Martha Warncock, MD) From Webb WR et al, 2009)

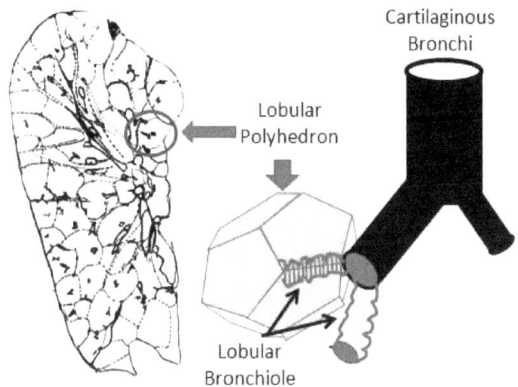

Figure 4: Lobular bronchiole/polyhedron interdependence

The left picture is a cross-sectional view of right upper lobe of human lung. Note that many lobular polyhedrons are aggregated in the lobe, and the bronchial tree is buried between the polyhedrons. A non-cartilaginous lobular bronchiole is adjoined to the edge of a lobular polyhedron by adventitial connective tissue. (modified from Min K et al,2012)

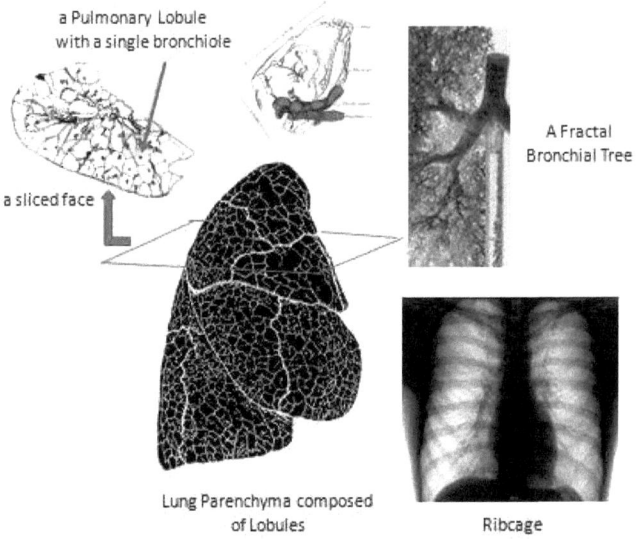

a Pulmonary Lobule
with a single bronchiole

A Fractal
Bronchial Tree

a sliced face

Lung Parenchyma composed
of Lobules

Ribcage

Figure 5: Components of Respiratory System

The ribcage consists of thoracic structures and the diaphragm, the right lung parenchyma consists of many lobules, a sliced face of right upper lobe lobules with a single bronchiole, and a fractal bronchial tree integrates many lobules. (modified from Min K. et al, 2013)

3.4 Patterns in Contractions of Lobular bronchioles during Breathing

The respiratory system consists of three functional components; ribcage, lung parenchyma, and bronchial tree (Figure 5). The ribcage consists of thoracic structures and the diaphragm as the basic driver of respiratory motion. The lung parenchyma consists of Miller's secondary lobules, each of which has a single bronchiole (Figure 3, 5). During spontaneous breathing depending on motions of ribcage, the fractal bronchial tree integrates many pulmonary lobules into ventilation through phasic shifted contractions of lobular smooth muscles. However, during a forced expiration the asynchronous phasic motions of lobular bronchioles would become simultaneous (Figure 6).

3.5 State Variables in Noisy Breathing

Spontaneous breathing is described as a series of tidal volumes or changes in respiratory rhythm. A series of tidal volumes is produced from the neural activity of the respiratory center in the brain. The neural activities of the respiratory center induce changes in the length of respiratory muscles, which are transformed into changes in the pleural pressure through the architectural properties of the ribcage. The changes in the pleural pressure are transformed to the alveolar pressure through the lung parenchyma. The alveolar pressure is transformed into airway pressure by the pulmonary lobule, and goes into the environment by producing airflows through the fractal bronchial tree (Figure 6A). It is important to note in Figure 6A that there are two origins of fluctuations in this process: in the respiratory rhythm generator (the neural center of respiration) and in the fractal airway modulator (the phasic asynchronous contractions of airway smooth muscles in the lobular bronchioles). Then, based on that bronchial flow F(t) is composed of N-number of phasic lobular flow(q), a tidal volume V_T is defined as following,

$$V_T = \int_0^\tau |F(t)|\,dt = \int_0^{\tau_I} \left(q \sum_j^N \delta_j\right)dt + \int_{\tau_I}^\tau \left(q \sum_j^N \delta_j\right)dt$$

$$= qN\{\tau_I \overline{\delta_I} + (\tau - \tau_I)\overline{\delta_E}\} \cdots (3.5.1)$$

where τ_I is inspiration priod, and δ_j is 0 or 1 for the j-th lobular bronchiole. $\overline{\delta_I}$ and $\overline{\delta_E}$ are the mean value of $\{\delta_j\}$ during inspiration and expiration, respectively. On steady state it is presumed that $\overline{\delta_I} = \overline{\delta_E} = \overline{\delta}$ and is less than 1 or $\overline{\delta} = \sin\theta$, then V_T is expressed by the following,

$$V_T = (qN\tau)\sin\theta \cdots (3.5.2)$$

During a voluntary forced expiration maneuver each lobule exhales a flow simultaneously (Figure 6B). Then, the forced expiration volume in one second ($FEV_{1.0}$) is defined by the following,

$$FEV_{1.0} = \int_0^1 F(t)\,dt = qN$$

Thus, the state variable of noisy breathing x is V_T normalized by $FEV_{1.0}$ as the following,

$$x = \frac{V_T}{FEV_{1.0}} = \tau\sin\theta \cdots (3.5.3)$$

The variable τ is the interbreath interval (IBI), and $\sin\theta$ is the proportion of simultaneously relaxed lobular bronchioles in the lung during a breath.

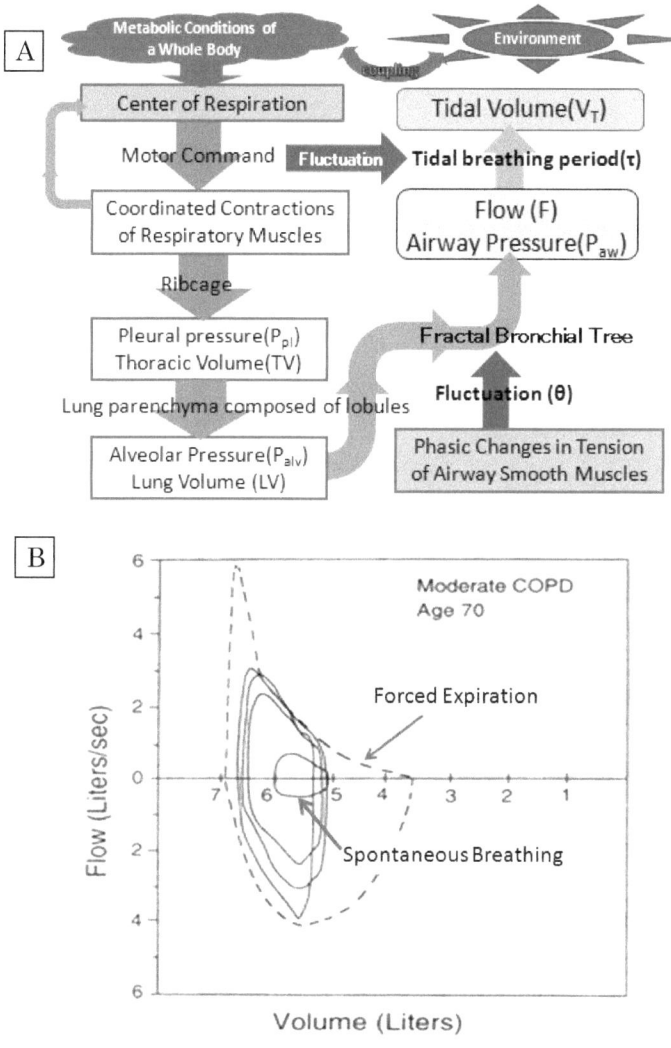

Figure 6: A: a series of tidal volumes is produced in the respiratory system. B: flow-volume trajectories of a human subject (70 year-old male of mild obstructive airway disease (COPD).
Two types of trajectories are seen: trajectories of spontaneous breathing and forced breathing.
(modified from Min K, 2013)

References

Kamiya, A., and Takahashi, T. Quantitative assessments of morphological and functional properties of biological trees based on their fractal nature. *2007 J Appl Physiol*, **102**: 2315-2323

Krahl VE. A method of studying the living lung in the closed thorax, and some preliminary observations. *1962 Angiology 14*: 149-159

MacDonald, N. Trees and networks in biological models. Part III. Branching structures: description, biophysics, and simulations. *1983 John Wiley and Sons, Chichester, New York, Brisbane, Toronto, Singapore.*

Mandelbrot, B.B. The fractal geometry of nature. *1983 Freeman, New York.*

Min K, Kawai M, Tamoto A, Mozai T, and Uchida E. Geometrical Analysis on the Pulmonary Lobular polyhedron (PLP) and Consideration of the way of arrangement of PLPs in the Lung. (Abstract in English) (Japanese) *1987 Jpn J Thoracic Dis* , **25**, 722-730

Min K , Hosoi K , Kinoshita Y , Hara S , Degami H , Takada T and Nakamura T Use of fractal geometry to propose a new mechanism of airway-parenchymal interdependence. *2012 Open Journal of Molecular and Integrative Physiology*, **2**, 14-20

Min, K. A Stochastic Optimal Control Theory to Model Spontaneous Breathing. *2013 Applied Mathematics*, **4**, 1537-1546

Suwa N and Takahashi T. Morphological and morphometrical analysis of circulation in hypertension and ischemic kidney. *1971 Urban and Schwarzenberg, München, Germany.*

Webb WR, Mueller NL, and Naidich DR. 2. Normal Lung Anatomy in High-Resolution CT of the Lung, 3rd edition *2009 Lippincott Williams & Wilkins*

Chapter 4 Stochastic State Model of Spontaneous Breathing and Optimal Control Theory

4.1 State model of Spontaneous breathing

The spontaneous breathing is characterized by a series of respiratory variables $\{V_T\}$. One will consider the series $\{V_T\}$ as a stochastic process $\{X(t)\}$ characterized by the following stochastic equation with the state variable x and the variance σ^2,

$$dX(t) = \begin{cases} b(x,t)dt + \sigma dw(t) & dt > 0 \\ b^*(x,t)dt + \sigma dw^*(t) & dt < 0 \end{cases} \cdots (4.1.1)$$

where w(t) and w*(t) are the forward Wiener process and the backward Wiener process with unit variance parameter, respectively. The function, b(x,t) or b*(x,t) is called the forward drift function or the backward drift function of state variable x at the given t, respectively as follows,

$$b(x,t) = \lim_{dt \to 0+} E_t\left[\frac{dX(t)}{dt}\right] = \lim_{dt \to 0+} E_t\left[\frac{X(t+dt) - X(t)}{dt}\right] \cdots (4.1.2a)$$

$$b^*(x,t) = \lim_{dt \to 0-} E_t\left[\frac{dX(t)}{dt}\right] = \lim_{dt \to 0-} E_t\left[\frac{X(t) - X(t-dt)}{dt}\right] \cdots (4.1.2b)$$

where $E_t [\,]$ denotes the conditional expectation of stochastic variables at the given t.

4.2 Optimal Controlled Conditions of the Stochastic State Model

Optimal control deals with the problem of finding a control law for a given system such that a certain optimality criterion is achieved. The optimality criterion includes a value of H similar to the total energy of a mechanical system. In the case of noisy breathing, a cost function H(t) should be of equilibrium at optimal controlled conditions as follows,

$$H(t) = E\left[\frac{1}{2}\left\{\frac{b(x,t)^2}{2} + \frac{b^*(x,t)^2}{2}\right\} + U(x)\right]$$

$$\frac{dH(t)}{dt} = 0$$

where $U(x)$ is a potential function of the respiratory system. By use of the probability density function $\rho(x,t)$, the stochastic optimal controlled conditions are expressed by the following,

$$\frac{d}{dt}\int\left[\frac{1}{2}\left\{\frac{b(x,t)^2}{2} + \frac{b^*(x,t)^2}{2}\right\} + U(x)\right]\rho(x,t)dx = 0 \cdots (4.2.1)$$

4.3 Einstein's Diffusion Equation

Consider a function f a continuous real valued function. The variable $f(X(t))$ is also a stochastic variable. Based on the definitions of stochastic differentials, two differentials for $f(X(t))$ are defined by use of the state variable x as follows,

$$\lim_{dt\to 0+} E_t\left[\frac{df(X(t))}{dt}\right] = \lim_{dt\to 0+} E_t\left[\frac{f(X(t+dt)) - f(X(t))}{dt}\right]$$

$$= b(x,t)\frac{df(x)}{dx} + \frac{\sigma^2}{2}\frac{d^2f(x)}{dx^2} \cdots (4.3.1a)$$

$$\lim_{dt\to 0-} E_t\left[\frac{df(X(t))}{dt}\right] = \lim_{dt\to 0-} E_t\left[\frac{f(X(t)) - f(X(t-dt))}{dt}\right]$$

$$= b^*(x,t)\frac{df(x)}{dx} - \frac{\sigma^2}{2}\frac{d^2f(x)}{dx^2} \cdots (4.3.1b)$$

Thus, the differential of $E[f(X(t))]$ by t is expressed as follows,

$$\frac{d}{dt}E[f(X(t))]$$

$$= \lim_{dt \to 0+} \frac{E[f(X(t+dt))] - E[f(X(t))]}{dt}$$

$$= \lim_{dt \to 0+} E\left[E_t\left[\frac{f(X(t+dt)) - f(X(t))}{dt}\right]\right]$$

$$= E\left[b(x,t)\frac{df(x)}{dx} + \frac{\sigma^2}{2}\frac{d^2f(x)}{dx^2}\right] \cdots (4.3.2)$$

When the series of stochastic variables $\{X(t)\}$ have a probability density function $\rho(x,t)$ of state variable x, the differential of $E[f(X(t))]$ by t is also expressed as follows,

$$\frac{d}{dt}E[f(X(t))] = \frac{d}{dt}\int f(x)\rho(x,t)dx = \int f(x)\frac{\partial\rho(x,t)}{\partial t}dx \cdots (4.3.3)$$

Comparing (4.3.2) and (4.3.3), the following relation is necessary if the function $f(x)$ is arbitrary,

$$\frac{\partial\rho(x,t)}{\partial t} = -\frac{d}{dx}(b(x,t)\rho(x,t)) + \frac{\sigma^2}{2}\frac{d^2}{dx^2}\rho(x,t) \cdots (4.3.4a)$$

Starting from (4.3.1b), the following equation is also necessary,

$$\frac{\partial\rho(x,t)}{\partial t} = -\frac{d}{dx}(b^*(x,t)\rho(x,t)) - \frac{\sigma^2}{2}\frac{d^2}{dx^2}\rho(x,t) \cdots (4.3.4b)$$

By combining (4.3.4a) and (4.3.4b), two equations are obtained as follows,

$$\frac{\partial\rho(x,t)}{\partial t} + \frac{d}{dx}\left\{\frac{1}{2}(b(x,t) + b^*(x,t))\right\} = 0 \cdots (4.3.5a)$$

$$\frac{d}{dx}\left\{\frac{1}{2}(b(x,t) - b^*(x,t))\rho(x,t)\right\} = \frac{\sigma^2}{2}\frac{d^2}{dx^2}\rho(x,t) \cdots (4.3.5b)$$

Here, let us introduce two functions, $v(x,t)$ and $u(x,t)$ as follows

$$v(x,t) = \frac{1}{2}\{b(x,t) + b^*(x,t)\} \cdots (4.3.6a)$$

$$u(x,t) = \frac{1}{2}\{b(x,t) - b^*(x,t)\} \cdots (4.3.6b)$$

Then, the functional relationships of $v(x,t), u(x,t)$, and $\rho(x,t)$ can be established by the following two equations,,

$$\frac{\partial \rho(x,t)}{\partial t} + \frac{d}{dx}\big(v(x,t)\rho(x,t)\big) = 0 \cdots (4.3.7a)$$

$$\frac{d}{dx}\big(u(x,t)\rho(x,t)\big) = \frac{\sigma^2}{2}\frac{d^2}{dx^2}\rho(x,t) \cdots (4.3.7b)$$

(4.3.7b) is equal to the diffusion equation of Einstein as follows

$$u(x,t) = \frac{\sigma^2}{2}\frac{d}{dx}\log\rho(x,t) \cdots (4.3.8)$$

References

K. Yasue. Quantum Mechanics and Optimal Stochastic Control Theory, *2007 Kaimei-shya, Tokyo*: Chapter 3 (in Japanese)

Chapter 5 Shrödinger's Wave Equation for Noisy Breathing

5.1. Shrödinger's Wave Equation as Optimal Controlled Conditions

According to (4.2.1) and (4.3.7a and 4.3.7b), the optimal condition of noisy breathing is defined using of functions $v(x,t), u(x,t)$ and $\rho(x,t)$, as follows

$$\frac{d}{dt}\int \left\{\frac{1}{2}\left(v(x,t)^2 + u(x,t)^2\right) + U(x)\right\}\rho(x,t)dx = 0 \ \cdots (5.1.1)$$

It is possible to transform (5.1.1) to the following equation (see Appendix),

$$\frac{\partial v(x,t)}{\partial t} = \left(u(x,t)\frac{d}{dx}u(x,t) + \frac{\sigma^2}{2}\frac{d^2}{dx^2}u(x,t)\right) - v(x,t)\frac{d}{dx}v(x,t)$$

$$- \frac{d}{dx}U(x) \ \cdots (5.1.2)$$

According to Einstein's diffusion equation of (4.3.8), the following equation is obtained,

$$\frac{\partial u(x,t)}{\partial t} = \frac{\sigma^2}{2}\frac{d}{dx}\frac{\partial \log\rho(x,t)}{\partial t} = \frac{\sigma^2}{2}\frac{d}{dx}\left\{\frac{\partial \rho(x,t)}{\partial t}\Big/\rho(x,t)\right\}$$

$$= -\frac{\sigma^2}{2}\frac{d}{dx}\left\{\frac{\frac{d}{dx}(v(x,t)\rho(x,t))}{\rho(x,t)}\right\} = -\frac{\sigma^2}{2}\frac{d}{dx}\left\{\frac{dv(x,t)}{dx} + v(x,t)\frac{\frac{d\rho(x,t)}{dx}}{\rho(x,t)}\right\}$$

$$= -\frac{\sigma^2}{2}\frac{d^2v(x,t)}{dx^2} - \frac{d}{dx}\left(v(x,t)u(x,t)\right) \ \cdots (5.1.3)$$

The probability density function $\rho(x,t)$ obeys the Fokker-Planck equation as follows,

$$\frac{\partial \rho(x,t)}{\partial t} = -\frac{d}{dx}\left(b(x,t)\rho(x,t)\right) + \frac{\sigma^2}{2}\frac{d^2\rho(x,t)}{dx^2} \ \cdots (5.1.4)$$

A set of transformations are applied to the functions $v(x,t)$ and $u(x,t)$ as follows,

$$v(x,t) = \sigma^2\frac{d}{dx}S(x,t)$$

$$u(x,t) = \frac{\sigma^2}{2}\frac{d}{dx}\log\rho(x,t)$$

Then, from (5.1.2), (5.1.3) and (5.1.4) a set of partial differential equations are obtained as follows,

$$\sigma^2\frac{\partial S(x,t)}{\partial t} = -U(x) - \frac{1}{2}\left[\sigma^2\frac{d}{dx}S(x,t)\right]^2 + \frac{\sigma^4}{2}\left(\frac{\frac{d^2}{dx^2}\sqrt{\rho(x,t)}}{\sqrt{\rho(x,t)}}\right) \cdots (5.1.5a)$$

$$\frac{\partial\rho(x,t)}{\partial t} = -\sigma^2\frac{d\rho(x,t)}{dx}\frac{dS(x,t)}{dx} - \sigma^2\rho(x,t)\frac{d^2S(x,t)}{dx^2} \cdots (5.1.5b)$$

Here let us introduce differential operators $\left(\nabla = \frac{d}{dx} \text{ and } \nabla^2 = \frac{d^2}{dx^2}\right)$ and a function $R(x,t) = \sqrt{\rho(x,t)}$, then the equations (5.1.5a) and (5.1.5b) are expressed in simpler equations as follows,

$$\sigma^2\frac{\partial S}{\partial t} + \frac{(\sigma^2\nabla S)^2}{2} + U - \frac{\sigma^4}{2}\frac{\nabla^2 R}{R} = 0 \cdots (5.1.6a)$$

$$\frac{\partial R^2}{\partial t} + \sigma^2\nabla\cdot(R^2\nabla S) = 0 \cdots (5.1.6b)$$

Let us introduce a complex function $\psi(x,t)$ as follows,

$$\psi = R\exp(iS/\sigma^2)$$

By use of ψ, the equations (5.1.6a) and (5.1.6b) can be transformed to a single motion equation which equates to Shrödinger's wave equation as follows,

$$-i\sigma^2\frac{\partial\psi}{\partial t} = -\frac{\sigma^4}{2}\frac{d^2\psi}{dx^2} + U\psi \cdots (5.1.7)$$

5.2. Distribution of Temporal and Regional Lung Ventilations

While noisy breathing is in the steady state of optimal controlled conditions, the cost function would be equal to an optimal value of H. Thus, the wave function of noisy ventilations is defined by the following

$$\left\{ -\frac{\sigma^4}{2}\frac{d^2}{dx^2} + U(x) \right\} \psi(x) = H\psi(x) \cdots (5.2.1)$$

When the state variable x is expressed by (3.5.3), the operator $\nabla^2 = \frac{d^2}{dx^2}$ is expressed by the following

$$\frac{d^2}{dx^2} = \frac{1}{\tau^2}\frac{\partial}{\partial\tau}\left(\tau^2 \frac{\partial}{\partial\tau} \right) + \frac{1}{\tau^2\sin\theta}\frac{\partial}{\partial\theta}\left(\sin\theta \frac{\partial}{\partial\theta} \right) \cdots (5.2.2)$$

If the potential function $U(x)$ is dependent on only the variable τ, (5.3.1) can be transformed to the equation (5.2.3) after rewriting the wave function as $\Psi(x)=T(\tau)Y(\theta)$,

$$\frac{1}{T(\tau)}\frac{\partial}{\partial t}\left(\tau^2 \frac{\partial T(\tau)}{\partial\tau} \right) - \frac{2\tau^2}{\sigma^4}(U(\tau) - H) = -\frac{1}{Y(\theta)}\left\{ \frac{1}{\sin\theta}\frac{\partial}{\partial\theta}\left(\sin\theta\frac{\partial Y(\theta)}{\partial\theta} \right) \right\} \cdots (5.2.3)$$

Each side term of (5.2.3) contain different single parameter τ or θ, thus each side term of (5.2.3) should be a constant χ. An equation for θ is obtained from the right side term of (5.2.3) as follows,

$$\frac{1}{\sin\theta}\frac{d}{d\theta}\left(\sin\theta\frac{dY(\theta)}{d\theta} \right) + \chi Y(\theta) = 0 \cdots (5.2.4)$$

When the transformation of $\cos\theta = s$ is applied to (5.2.4), the following equation is obtained,

$$(1 - s^2)\frac{d^2 Y(s)}{ds^2} - 2s\frac{dY(s)}{ds} + \chi Y(s) = 0 \cdots (5.2.5)$$

(5.2.5) is a Legendre equation, whose solutions are obtained as Legendre orthogonal polynomials only when $\chi = k(k + 1)$, where $k = 0,1,2, \cdots$ as follows,

$$Y(\theta) = Y_k(s) = \sum_{j=0}^{[k/2]} \frac{(-1)^j(2k - 2j)!}{2^k j! (k - j)! (k - 2j)!} s^{k-2j} \cdots (5.2.6)$$

That is, each solution is dependent on k as follows, $Y_0(s) = 1, Y_1(s) = s, Y_2(s) = (3s^2 - 1)/2$, $Y_3(s) = (5s^3 - 3s)/2$, and so on. One can obtain the probability density function by $\rho(s) = |Y_k(s)|^2$ with $\int_0^1 \rho(s)ds = 1$ as shown in Figure 7. It has been suggested that $\rho(s)$ would relate to patterns of temporal and regional ventilations emerging as a result of phasic contractions of smooth muscles in the lobular bronchioles. The parameter χ would be a marker for emerging pattern of regional ventilations in the lung. Venegas et al. recently demonstrated possibility of making images of temporal and regional distribution of pulmonary ventilation by the images of positron emission tomography (PET) (Figure 8).

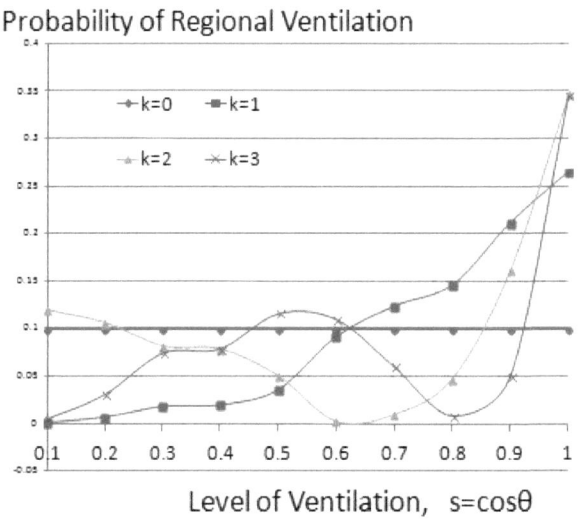

Figure 7: Probability Distribution of Regional Ventilations in the Lung
Each distribution density $\rho(s)$ was obtained by $\rho(s) = |Y_k(s)|^2$, where $Y_0(s) = 1, Y_1(s) = s, Y_2(s) = (3s^2 - 1)/2$, or $Y_3(s) = (5s^3 - 3s)/2$. The probability of regional ventilations was calculated by $P(s) = \int_{s-0.1}^{s} \rho(t)dt$. Note that the distribution pattern of regional ventilations is quite different according to the parameter k.

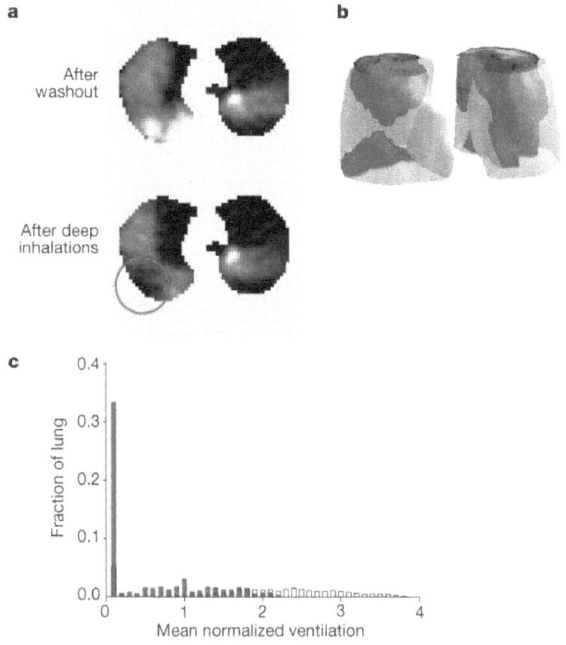

Figure 8: Heterogeneity in bronchoconstriction of an asthmatic's lung
(from Venegas G et al,2005)

a, Residual intrapulmonary [13]NN tracer gas activity in a representative lung cross-section after intravenous bolus injection of [13]NN-saline solution. Tracer concentration increases according to the following colour scale: black (no tracer), red, yellow, and white (highest). The insoluble tracer is washed out during breathing or was retained inside large ventilation defects. After deep inhalations (lower panel), tracer clearance is enhanced from parts of these defects (circle). **b**, Volumetric rendering of ventilation defects (red) and the external surface of the lungs (blue); image orientation is as if the subject were standing facing the reader. **c**, Histograms of mean normalized regional lung ventilation across differentiating units inside (red) and outside (blue) of the ventilation defects.

5.3. Shrödinger's Wave Equation for Inter-Breath Intervals

From the left side term of (5.2.3) an equation is obtained as follows,

$$\frac{1}{\tau^2}\frac{d}{d\tau}\left(\tau^2\frac{dT(\tau)}{d\tau}\right) + \frac{2\tau^2}{\sigma^4}\left(H - U(\tau)\right)T(\tau) - \frac{\chi}{\tau^2}T(\tau) = 0$$

When $P(\tau)$ is introduced as $P(\tau) = \tau T(\tau)$, the following equation is obtained as another wave equation for $P(\tau)$,

$$\left(-\sigma^4\frac{d^2}{d\tau^2} + U(\tau) + \frac{\sigma^4\chi}{2\tau^2}\right)P(\tau) = HP(\tau) \cdots (5.3.1)$$

One can produce a distribution density by $\rho(\tau) = |P(\tau)|^2$ at the optimal value of H, which is a probability of inter-breath intervals (IBIs) observed between τ and $\tau + d\tau$.

For an optimal condition of H, one assumes that the wave function $P(\tau)$ is expressed by two functions $\Phi(\tau)$ and $\Psi(\tau)$ as follows,

$$-\sigma^4\frac{d^2\Phi(\tau)}{d\tau^2} + \left(U(\tau) + \frac{\sigma^4\chi}{2\tau^2}\right)\Phi(\tau) = H\Phi(\tau) \cdots (5.3.2a)$$

$$-\sigma^4\frac{d^2\Psi(\tau)}{d\tau^2} + \left(U(\tau) + \frac{\sigma^4\chi}{2\tau^2}\right)\Psi(\tau) = H\Psi(\tau) \cdots (5.3.2b)$$

By calculating $(5.3.2a) \times \Psi(\tau) - (5.3.2b) \times \Phi(\tau)$, one obtains the following equation:

$$\frac{d}{d\tau}\left[\Phi(\tau)\frac{d\Psi(\tau)}{d\tau} - \Psi(\tau)\frac{d\Phi(\tau)}{d\tau}\right] = 0 \cdots (5.3.3)$$

If $\tau \to \infty$, then both $\Psi(\tau)$ and $\Phi(\tau) \to 0$. Therefore, (5.3.3) is transformed to (5.3.4) as follows,

$$\Phi(\tau)\frac{d\Psi(\tau)}{d\tau} = \Psi(\tau)\frac{d\Phi(\tau)}{d\tau} \quad \text{or} \quad \Psi(\tau) \propto \Phi(\tau) \cdots (5.3.4)$$

Thus, the state of the rhythm generator is uniquely determined with dependence on the value of H.

5.4. How can Probability Density Functions of IBIs characterize the function of the Central Rhythm Generator?

The probability density function ρ(IBI) is expressed by the wave function in (5.3.1) as follows: $\rho(\text{IBI}) = |P(\tau)|^2$. When the wave function $P(\tau)$ is expressed by the following,

$$P(\tau) = \exp\big(-f(\tau)\big) \cdots (5.4.1)$$

$U(\tau)$ is expressed by the following,

$$U(\tau) = H - \sigma^4 \left[\frac{d^2 f(\tau)}{d\tau^2} - \left(\frac{df(\tau)}{d\tau}\right)^2 + \frac{\chi}{2\tau^2} \right] \cdots (5.4.2)$$

The equation (5.4.2) explains how the probability density function (PDF) relates to the function of the central rhythm generator.

Respiratory rhythm generation arises in the medullary neurons that initiate rhythmic inspiratory and expiratory activity. Several studies suggest that the pre-Bötzinger complex, a discrete group of propriobulbar neurons in the ventrolateral medulla, plays a critical role in respiration rhythm generation, although this hypothesis is not without controversy. Pattern-forming neurons include premotoneurons and motoneurons in the brain stem or spinal cord, where complex activation patterns arise from interactions between their intrinsic properties and synaptic inputs. Pattern formation establishes the detailed spatio-temporal motor output of respiratory muscles, coordinating their activation to produce a breath with the appropriate characteristics. These coordinated, complex interactions among groups of neurons in the brain produce an optimal breathing rhythm which is described in the stochastic model by $P(\tau)$ as shown by the equation (5.4.2).

Mitchell and Johnson have stated that a comprehensive conceptual framework of neuroplasticity in the respiratory control system is lacking. However, the equation (5.4.2) can provide a comprehensive framework for respiratory rhythm generation since this expression includes an optimal total energy H of the respiratory system, the topographical distribution parameter χ of regional ventilation in the lung, and the probability density function (PDF) of inter-breath intervals (IBIs). Frey et al. demonstrated the fractal properties of PDFs of IBIs in preterm, term babies and and Fadel et al. a third of adults at rest (Figure 9). When there are fractal properties in PDFs of IBIs as follows $\rho(\tau) \propto \tau^{-\alpha}$, according to (5.4.2) the potential $U(\tau)$ of the RRG is expressed by the following,

$$U(\tau) = H - \frac{\sigma^4}{2\tau^2}[-2\alpha(\alpha+1) + \chi] \cdots (5.3.1)$$

This potential of the RRG shows that development of the RRG in infants leads to a change in parameters α and χ, but no change in the structure of the potential function $U(\tau)$. If a change in the structure of the potential function signals neuroplasticity, then the developmental change of the RRG is not a neuroplastic process.

I will summarize this chapter: variability in spontaneous breathing is not simply attributable to random noise superimposed on a regular respiratory process. Respiratory variability originates from RRG and airway smooth muscles (ASM), thus the state variable of respiratory motion was able to be described by a polar variable consisting of both inter-breath interval and pattern of contractions of ASM. Optimal control theory on a stochastic state model for spontaneous noisy breathing could produce a Shrödinger's wave equation as the motion equation, from which two differential equations were obtained: one for the RRG and another for the modulator of ASM in the lung. From these equations, the function of RRG was defined as a complex function including probability density functions in both rhythm and amplitude of spontaneous noisy breathing.

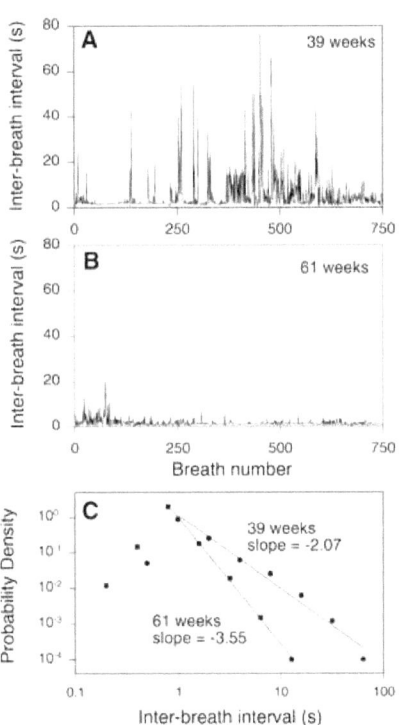

Figure 9. Breathing irregularities in a preterm infant

Interbreath intervals as a function of breath number at postconceptional ages of 39 (*A*) and 61 weeks (*B*). (*C*) Probability density distributions estimated from *A* and *B* and straight line fits on a log–log graph (from Frey et al. 2004).

References

Frey U, Silverman M, Barrabasi AL, Suki B. Irregularities and power law distributions in the breathing pattern in preterm and term infants. *1998 J Appl Physiol 85*: 789-797

Killingbeck J. Microcomputer Quantum Mechanics, *1983 Adam Hilger Ltd*

Min, K. A Stochastic Optimal Control Theory to Model Spontaneous Breathing. *2013 Applied Mathematics*, **4**, 1537-1546

Toda M. Quantum Mechanics 30 Lectures. *1999 Asakura Ltd, Tokyo* (Japanese)

Yasue K. Quantum Mechanics and Optimal Stochastic Control Theory, *2007 Kaimei-shya, Tokyo*: Chapter 3 (in Japanese)

Chapter 6: Again, what is the Biological Variability?- as a Conclusion

6.1 Origin of biological variability

In the stochastic process model I have proposed in this monograph, fluctuations in respiratory state variables have been described by the stochastic differential equation on the basis of variance parameter σ^2. Until today, little work has been done in regard to the origin of variance parameter σ^2. According to recent experimental studies *in vitro*, Suki and colleagues have suggested that energetic and metabolic fluctuations at the level of the cell are essential components of biological variability. If any biological variability comes from the fluctuation of energy at the level of cell, it will be acceptable that fluctuations of physiological variables are described on the basis of variance parameter σ^2. In other words, it is the hypothesis that a cell can produce the variance parameter σ^2 from molecular fluctuations through own metabolic processes. Thus, the biological variability at the organ level would also be described on the basis of variance parameter σ^2 since an organ is organized cluster of cells.

6.2 Potential of Biological Variability

The stochastic control model analysis described in this monograph will provide a new concept of biological processes beyond respiratory fluctuations. Let us pay attention to a set of equations (6.2.1) and (6.2.2), which is rewritten from equations (5.1.6a) and (5.1.6b) by replacing $\sigma^2 S$ with S as follows,

$$\frac{\partial S}{\partial t} + \frac{(\nabla S)^2}{2} + U - \frac{\sigma^4}{2}\frac{\nabla^2 R}{R} = 0 \cdots (6.2.1)$$

$$\frac{\partial R^2}{\partial t} + \nabla \cdot (R^2 \nabla S) = 0 \cdots (6.2.2)$$

When the term $-\sigma^4 \nabla^2 R/2R$ is very smaller compared with the term $(\nabla S)^2/2$, the following equation is obtained,

$$\frac{\partial S_c}{\partial t} + \frac{(\nabla S_c)^2}{2} + U = 0 \cdots (6.2.3)$$

In the above $S=S_c$ is written to indicate that we are dealing with normal processes without noise. This expression is well known as the classical Hamilton-Jacobi equation representing a classical process with momentum,

$$v(x, t) = \nabla S_c \cdots (6.2.4)$$

Thus, it is note that the equation (6.2.1) for noisy processes differs from the equation (6.2.3) for normal processes only by the term $-\sigma^4 \nabla^2 R/2R$ which evidently can be regarded as playing the role of additional potential in which we may call processes with biological variability. Thus, I will introduce a new potential for biological variability;

$$BV = -\frac{\sigma^4}{2} \frac{\nabla^2 R}{R} \cdots (6.2.5)$$

A Hamilton-Jacobi equation is obtained for biological noisy processes as follows,

$$\frac{\partial S}{\partial t} + \frac{(\nabla S)^2}{2} + U + BV = 0 \cdots (6.2.6)$$

The equation (6.2.2) expresses the conservation of probability, but for an ensemble of biological processes which satisfies (6.2.6) rather than (6.2.3). This is a Bohmian equation of quantum mechanics. Regarding to Bohm's ontological interpretation of quantum mechanics, I would like to summarize the biological processes with fluctuation as follows,

1. Biological systems are composed of a large number of cells with variance parameter σ^2.

2. A cell is never separate from the potential field of biological variability (the potential BV) that fundamentally affects it. This field is given by two functions R

and S, or alternatively by the wave function $\Psi = R\exp(iS/\sigma^2)$. Ψ satisfies Shrödinger's wave equation.

3. A cell has an equation of motion for changing state variable as follows,

$$\frac{dv}{dt} = -\nabla(U) - \nabla(BV) \cdots (6.2.7)$$

This means that the forces inducing a change in the state of cell are not only the classical potential $-\nabla(U)$, but also the new potential $-\nabla(BV)$.

4. The velocity of change in state variables of a cell is restricted to $v = \nabla S$.

5. In a statistical ensemble of cells, selected so that all have the same BV field, the probability density is R^2. If R^2 holds initially, the conservation equation (6.2.2) guarantees that it will hold for all time. Thus, when we analyze the organ consisting of many cells, biological processes at the organ level are explained as a statistical ensemble of biological processes of cells with a same wave function.

6.3 A Proposal as Conclusion

Respiratory variables, including tidal volume and respiratory rate, display significant variability. The probability density function (PDF) of respiratory variables has been shown to contain clinical information and can predict the risk for exacerbation in asthma. However, it is uncertain why this PDF plays a major role in predicting the dynamic conditions of the respiratory system. I have introduced a stochastic optimal control model for noisy spontaneous breathing, and obtains a Shrödinger's wave equation as the motion equation that can produce a PDF as a solution. Based on the fractal bronchial tree model, the tidal volume variable was expressed by a polar coordinate, by use of which the Shrödinger's wave equation of inter-breath intervals (IBIs) was obtained. Through the wave equation of IBIs, the respiratory rhythm generator was characterized by the potential function including the PDF and the parameter concerning the topographical distribution of regional pulmonary ventilations. The stochastic model was assumed to have a common variance parameter in the state variables, which would originate from the variability in metabolic energy at the cell level. Thus, I have proposed a new insight on biological variability by the potential field (BV) according to Bohm's ontological interpretation of quantum mechanics. As a conclusion, I would like to give a proposal that an integrated function of biological organ would appear by establishing a common field of BV among cells, and that it is necessary to measure biological variability in order to understand what a living organ is.

References

Bohm D and Hiley BJ. The Undivided Universe *1992 Routledge, NY*

Suki B, Martinez N, Parameswaran H, Majumdar A, Dellaca R, Berry C, Pillow J J, and Bartolak-Suki E. Variability In The Respiratory System: Possible Origins And Implications, *2012ATS, B29 THE LUNG ON THE BORDER BETWEEN ORDER AND CHAOS*: A2682

$$\frac{d}{dt}\int \left\{\frac{1}{2}(v(x,t)^2 + u(x,t)^2) + U(x)\right\}\rho(x,t)dx = 0 \cdots (5.1.1)$$

The first term of (5.1.1) is calculated as follows,

$$\frac{d}{dt}\int \frac{1}{2}v(x,t)^2\rho(x,t)dx = \int \frac{1}{2}\frac{\partial v(x,t)^2\rho(x,t)}{\partial t}dx$$

$$= \int \left\{v(x,t)\frac{\partial v(x,t)}{\partial t}\rho(x,t) + \frac{1}{2}v(x,t)^2\frac{\partial \rho(x,t)}{\partial t}\right\}dx$$

$$= \int \left\{v(x,t)\frac{\partial v(x,t)}{\partial t}\rho(x,t) - \frac{1}{2}v(x,t)^2\frac{d^2}{dx^2}(v(x,t)\rho(x,t))\right\}dx$$

$$= \int \left\{\frac{\partial v(x,t)}{\partial t} + v(x,t)\frac{\partial v(x,t)}{\partial t}\right\}v(x,t)\rho(x,t)dx \cdots (A.1)$$

The second term of (5.1.1) is also transformed as follows,

$$\frac{d}{dt}\int \frac{1}{2}u(x,t)^2\rho(x,t)dx = \int \frac{1}{2}\frac{\partial u(x,t)^2\rho(x,t)}{\partial t}dx$$

$$= \int \left\{u(x,t)\frac{\partial u(x,t)}{\partial t}\rho(x,t) + \frac{1}{2}u(x,t)^2\frac{\partial \rho(x,t)}{\partial t}\right\}dx$$

$$= \int \left\{\frac{\sigma^2}{2}u(x,t)\frac{\partial}{\partial t}\left(\frac{d}{dx}\log\rho(x,t)\right)\rho(x,t)\right.$$

$$\left. -\frac{1}{2}u(x,t)^2\frac{d}{dx}(u(x,t)\rho(x,t))\right\}dx$$

$$= \int \left\{ \frac{\sigma^2}{2} u(x,t) \frac{\frac{d\rho(x,t)}{dx}}{\rho(x,t)^2} \frac{d}{dx} \left(v(x,t)\rho(x,t) \right) \rho(x,t) - \frac{\sigma^2}{2} u(x,t) \right.$$

$$\left. \cdot \frac{d}{dx} \left(\frac{dv(x,t)}{dx} \rho(x,t) \right) - \frac{1}{2} u(x,t)^2 \frac{d}{dx} \left(v(x,t)\rho(x,t) \right) \right\} dx$$

$$= \int \left\{ \frac{1}{2} u(x,t)^2 \frac{d}{dx} \left(v(x,t)\rho(x,t) \right) - \frac{\sigma^2}{2} u(x,t) \frac{d^2}{dx^2} \left(v(x,t)\rho(x,t) \right) \right\} dx$$

$$= - \int \left\{ u(x,t) \frac{d}{dx} u(x,t) + \frac{\sigma^2}{2} \frac{d^2 u(x,t)}{dx^2} \right\} (v(x,t)\rho(x,t)) dx \quad \cdots \text{(A. 2)}$$

The third term of (5.1.1) is expressed by following,

$$\frac{d}{dt} \int U(x)\rho(x,t)dx = \int U(x) \frac{\partial\rho(x,t)}{\partial t} dx$$

$$= - \int U(x) \frac{d}{dx} \left(v(x,t)\rho(x,t) \right) dx$$

$$= \int \frac{dU(x)}{dx} \left(v(x,t)\rho(x,t) \right) dx \quad \cdots \text{(A. 3)}$$

By combining (A.1), (A.2) and (A.3), the criterion of optimal control is expressed by the following,

$$\frac{d}{dx} \int \left\{ \frac{1}{2} (v(x,t)^2 + u(x,t)^2) + U(x) \right\} \rho(x,t)dx$$

$$= \int \left[\frac{\partial v(x,t)}{\partial t} - \left(u(x,t) \frac{du(x,t)}{dx} + \frac{\sigma^2}{2} \frac{d^2 u(x,t)}{dx^2} \right) + v(x,t) \frac{dv(x,t)}{dx} \right.$$

$$\left. + \frac{dU(x)}{dx} \right] v(x,t)\rho(x,t)dx = 0 \quad \cdots \text{(A. 4)}$$

The equation (5.1.2) is obtained as the necessity for the criterion of control (A.4) as follows,

$$\left\{\frac{\partial v(x,t)}{\partial t} - \left(u(x,t)\frac{du(x,t)}{dx} + \frac{\sigma^2}{2}\frac{d^2u(x,t)}{dx^2}\right) + v(x,t)\frac{dv(x,t)}{dx}\right\} + \frac{dU(x)}{dx} = 0$$

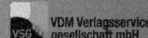

Printed by Books on Demand GmbH, Norderstedt / Germany